Table of Contents

Chapter One - Key Words Explained 4

Chapter Two – Explaining the Five Diseases (in Question and Answer Format) 6

Section One: Cystic Fibrosis - A Genetic Disease 6

Section Two: Malaria - An Infectious Disease Spread by a Vector .. 14

Section Three: AIDS - A Viral Infectious Disease 19

Section Four: Cholera - A Water-Borne Bacterial Infectious Disease 23

> As the main symptoms of diarrhoea and vomiting can cause severe dehydration; which itself can lead to shock due to low blood pressure; so the most common treatment is oral re-hydration therapy (ORT), which helps to replace lost fluid and salts. About eighty percent of those given ORT recover successfully from cholera. 24
>
> However, severely cases may need intravenous fluids and antibiotics to reduce diarrhea and kill the infecting cholera bacteria. 24

Section Five: Tuberculosis (TB), a Droplet-Infection Bacterial Disease 26

Chapter Three: Exam Questions and Answers 31

Understanding Five Major Worldwide Diseases; AIDS, Malaria, Cholera, Tuberculosis and Cystic Fibrosis

Questions and Answers Which Help You Understand the Causes. Symptoms and Cures

Why I wrote this book

Since the epidemic began in the early 1980s it is estimated that over thirty million people worldwide have died from AIDS (the illness caused by the HIV virus). Around the same number are currently living with the disease and two million more people a year are becoming infected with it.

Over three hundred million cases of malaria occur each year worldwide, with around one million people a year dying of it.

More than three million people a year worldwide get cholera with around one hundred thousand people dying from it per year.

Worldwide, more than one million people a year die of tuberculosis (TB) and around ten million people a year become newly infected with it.

More than sixty thousand people worldwide suffer from cystic fibrosis. People with the disease on average, tend to die before they reach forty years old, sadly.

This book is about explaining what causes these diseases and what cures or treatments are available.

As they represent different types of disease in each case; I believe someone reading this will gain the understanding of not only AIDS, malaria, cholera, TB and cystic fibrosis; but also many other diseases that have the same or similar basis.

This book will particularly benefit students of biology and medicine; but is written in understandable language for non-students; who may just have an interest in these diseases.

I hope that by reading this book you gain an insight into the fascinating subject of the biology of these illnesses and that what you learn will benefit you and others around you.

Chapter One - Key Words Explained

Cystic Fibrosis – is an inherited, genetic disease. Poor lung and pancreas function is the result of a mutant gene that produces a non-functional protein.

AIDS (Acquired Immune Deficiency Syndrome) – is the name for the disease and is caused by poor functioning of the immune system due to infection by the HIV virus.

HIV (Human Immunodeficiency Virus) - infects the T cells of the immune system and destroys them, leading to AIDS. Infection with HIV does not always lead to AIDS developing immediately, it can take many years.

Cholera – an infectious disease that is caused by bacteria, 'Vibrio cholerae'. It is usually spread through a contaminated water supply.

Malaria – an infectious disease caused by a protozoan that is spread by a vector – the mosquito.

Tuberculosis (TB) – infectious bacterial disease that mainly affects the lungs. It is caused by Mycobacterium tuberculosis (also M.Bovis is places where milk is not sterilised).
M. tuberculosis is transmitted from person to person by droplet infection when a sufferer coughs or sneezes.

Infectious disease – a disease caused by an infection by a pathogenic microbe, such as a bacterium, virus, protozoan, fungi or multi-cellular parasite.

Deficiency Disease – a disease caused by a lack of a certain nutrient in the diet e.g. blindness caused by a lack of vitamin A.

Inherited Disease – a disease caused by genes inherited from one or both of the parents e.g. cystic fibrosis is caused by a recessive allele that is inherited from both parents.

Mental Disease – affects the mind, e.g. schizophrenia or depression.

Self-inflicted Disease – a disease caused by the person own behaviour, usually involving abuse of drugs such as alcohol and nicotine. E.g. cirrhosis of the liver is often cause by alcohol abuse.

Cystic Fibrosis- this is a genetic illness caused by a faulty allele for a membrane transport protein in goblet cells in the epithelium of the bronchioles. The malfunction of this protein causes a build-up of mucus in the lungs leading to chronic lung infections which may damage the lungs. Build-up of mucus in the pancreatic duct may also lead to poor digestion.

Risk Factor – something that predisposes you to get a disease –e.g. being obese can predispose you to having CHD.

Vector – something that carries a disease from one person to another e.g. the mosquito in the case of malaria.

Mutant Gene - a mutant gene is a gene that is different from its normal, working form.

Chapter Two – Explaining the Five Diseases (in Question and Answer Format)

Section One: Cystic Fibrosis - A Genetic Disease

What is Cystic Fibrosis (CF)?

Cystic fibrosis is a genetic disorder that affects the respiratory, digestive and reproductive systems due to the production of abnormally thick mucus, and can lead to fatal lung infections.

The disease can also result in obstruction of the pancreas, hindering digestion and can prevent sperm moving through the sperm tubes, leading to infertility.

What Are The Symptoms of Cystic Fibrosis?

Symptoms usually start in early childhood and include:

- persistent cough (caused by mucus building up in the lungs)
- recurring chest and lung infections (also due to mucus build up in the lungs)
- a good appetite but with poor weight gain (caused by poor digestion due to blockage of the pancreatic duct by mucus)
- Shortness of breath (due to mucus blocking the passage of air to and from the lungs)
- Another sign is unusually salty sweat, which you might notice, for example, when you kiss your child.

How Can We Test For CF?

The sweat test is used as a standard diagnostic test for CF. This simply measures the amount of salt in the sweat with a high salt level indicating CF. Pulmonary function tests may show that breathing is compromised. Also, a chest x-ray may suggest the diagnosis. Relatives other than the parents of a child with cystic fibrosis may want to know if they're likely to have children with the disease. Genetic testing on a small blood sample can help determine who has a defective cystic fibrosis allele. During pregnancy, an accurate diagnosis of cystic fibrosis in the foetus is usually possible.

Is There A Genetic Test For CF?

Cystic fibrosis can now be confirmed through a genetic screening test, often before symptoms appear.

Is CF Always A Serious Disease?

The severity of cystic fibrosis varies greatly from person to person, even independent of their ages. This is partly because there are many different mutations of the gene responsible for causing CF. The severity of the disease is determined largely by how much the lungs are affected by it.

However, some deterioration in lung condition is inevitable, usually leading to debility and eventually death over a longish time period.

Adult sufferers may experience additional health challenges to lung deterioration including CF-related diabetes and osteoporosis.

How Long Can CF Sufferers Live For?

The good news is that the prospects of life expectancy for CF sufferers have improved steadily over the past few decades. This is mainly because treatments can now postpone some of the changes that occur in the lungs.

While thirty years ago few CF sufferers survived into adulthood, currently around four in ten of the CF sufferers are aged 18 and older. Also, many CF sufferers now live into their fifties and even sixties. The average age of survival is about thirty three years currently. Males and those without pancreatic problems tend to survive a little longer than others.

Gene therapy may greatly help in the future in treating CF.

How Might Gene Therapy Help Treat CF In The Future?

This treatment may help greatly to improve the lung condition and function of sufferers of CF. CF symptoms are due to a mutant form (allele) of a gene, so gene therapy aims to deliver millions of the healthy form of the gene to the CF sufferer's lungs.

If this is successful, the lung cells may use this healthy form of the gene rather than the ill form, helping to relieve symptoms and prevent lung damage and destruction.

The genes could be delivered by an inhaler (rather like the one used by asthma sufferers) to the lungs.

There are though a few problems with this gene therapy. One problem is that it has to be maintained continually, as the lung cells which take up the healthy gene will die and are replaced with new cells with the ill, mutant gene. A second problem is the unpredictable nature of what might happen to the healthy gene when it is delivered – particularly where it will do in the existing DNA – there is a fear it may disrupt existing DNA and cause mutation and even cancer.

Can CF Sufferers Live A Normal Life?

Yes they can lead a normal life in most cases. In spite of the difficulties of the disease, people with cystic fibrosis usually study or work normally until shortly before death.

Can CF Sufferers Have Children?

CF can cause reproductive problems - more than nine out of ten of men with CF are sterile. But, with new reproductive technologies, some male CF sufferers are becoming fathers.

Although many women with CF are able to conceive, limited lung function and other health factors may make it difficult to carry a child to term.

How Common Is CF?

There are around thirty thousand cystic fibrosis sufferers in the USA and about nine thousand cystic fibrosis sufferers in the UK.

CF is most common in people of white Northern European ancestry.

About four out of every ten thousand babies born in the UK have CF.

What Causes CF?

Genetic diseases are caused by the inheritance by the child, from the parents, of types of a gene (called 'alleles') which do not function properly because they are mutant (different).

There are thought to be hundreds of different mutations of the CF gene, some of which have been identified and sequenced by researchers.

In the case of cystic fibrosis this allele is recessive, so the child must have both alleles (one inherited from each parent) in order to suffer from the disease.

In order to have a child with CF then, both parents must therefore be carriers of the disease (it is unlikely that a sufferer of cystic fibrosis will be able to reproduce) - in other words the parents each have one allele for the disease (and one normal allele) but do not suffer any symptoms at all because the normal allele is dominant.

So, if both parents are carriers they have a 1 in 4 (25%) probability of having a child with cystic fibrosis every time they reproduce.
This is shown in the diagram below:

C - Normal (dominant) allele: c - cystic fibrosis (recessive) allele

Parents' genotypes: Cc   and   Cc

Gametes (sperm or egg):   C   c   C   c

When sperm and egg fuse they form the following children's genotypes: CC (25%); Cc (50%); cc (25%):

CC is normal; Cc is a carrier and cc has cystic fibrosis.

This means that each time two carriers of the disease have a child there is a 25% chance of having a child with cystic fibrosis; a 50% chance that the child will be a carrier of the cystic fibrosis gene (but be physically healthy); and a 25% chance that the child will not be carrier (have two healthy forms of the gene). As the allele is recessive, those children who are carriers do not have the disease and have absolutely no physical sign of it – a genetic test can however, reveal that they are carriers.

How Does A Faulty Allele Cause CF?

Genes code for proteins.

A genetic disease is caused by someone inheriting from their parents a mutant, faulty allele (of a gene) that does not code properly for the protein because it has an incorrect sequence of DNA codes; which result in a faulty protein with an incorrect sequence of amino acids. That protein as a consequence does not function correctly.

In the case of cystic fibrosis, it is a cell membrane protein called CFTR (that transports chloride ions) that does not function correctly when coded for by the cystic fibrosis allele.

The normal CTFR protein pumps chloride ions out of the cells lining the bronchioles in the lungs and the pancreatic duct, causing water to be moved by osmosis into the mucus that is produced there, making the mucus thin in consistency. This thin consistency is important in the lungs as it means the mucus can be removed by beating of the cilia in the ciliated epithelium layer that lines the inside of the bronchiole tubes – thereby cleaning the lungs. In the pancreatic duct the thin consistency of the mucus means it thinly and evenly coats the epithelium layer of cells there, stopping it from being digested by enzymes in solution that flow through it.

When someone has cystic fibrosis the CFTR protein does not function properly so there is not enough water in the mucus, causing it to have a thick consistency. This leads to the build up of this thick mucus in the bronchioles, because the ciliated epithelial cells cannot beat this mucus out of the lungs. The same build up of thick mucus happens in the pancreatic duct, eventually blocking it.

How does this build up of mucus in the lungs and pancreas cause the symptoms of cystic fibrosis?

This build up of mucus in the lungs causes the symptoms of coughing (which is a reflex action to try and remove the mucus) and chronic lung infections (as the mucus cannot be removed, it leads to the accumulation of bacteria and viruses which are trapped in it).

The build up of mucus in the pancreatic duct causes poor digestion because it prevents digestive enzymes reaching the small intestine.

Is There A Cure For Cystic Fibrosis?

There is no cure as it is an inherited genetic disease. However it is treatable with some success.

How Is Cystic Fibrosis Treated?

As CF is a genetic disease it is not curable (except possibly by lung and pancreas transplant) so treatment can only be designed to make CF easier to live with and to help prevent the damage of infections and mucus blocking the lungs and pancreatic duct.

There are a number of treatments that cystic fibrosis sufferers undergo; often on a daily basis.

These include; physiotherapy to try and clear the thick mucus out from the lungs.

Drugs: some to help thin the mucus and antibiotics help to fight the lung infections that tend to occur. Others, called bronchodilators, make it easier for air to enter and leave the lungs by increasing the size of the bronchiole tubes in the lungs.

Often enzyme pills are taken with food, to replace those lost due to the blockage of the pancreatic duct.

CF sufferers can also have a lung transplant, with in America, around two hundred of them taking this option every year. A lung transplant is usually necessary if the lungs have been damaged to a large extent by the infections due to CF – a pancreas transplant is also possible in sufferers whose pancreas has been damaged.

Section Two: Malaria - An Infectious Disease Spread by a Vector

What is the historical background to malaria?

Malaria has been noted for more than 4,000 years. From the Italian for "bad air," malaria has probably had a large influence on human history. The symptoms of malaria were described in ancient Chinese medical writings dating from 2700 BC.

Malaria became widely recognized in Greece by the 4th century BCE and Hippocrates noted the symptoms.

What Causes Malaria?

Malaria is a potentially life-threatening tropical disease caused by a tiny parasite called a 'protozoan'; specifically the Plasmodium protozoan; which can be transmitted from infected people to healthy people, through the bites of female mosquitoes.

How Does Someone Catch Malaria?

When the mosquito bites an infected person it takes up, in its mouthparts, spores from the protozoan, present in the infected person's blood. Most commonly it is a female of the type of mosquito called Anopheles, which bites at night; that transmits malaria like this.

When this mosquito bites an uninfected person it transfers the spores from its mouthparts to the person's blood, causing them to be infected with malaria.

In biological terms the Anopheles mosquito is called a 'vector' as it transfers the disease between people. It is thought that a single bite from an infected mosquito is sufficient to get malaria.

The protozoan spores then mature and multiply in the newly infected person's body and attack first the liver and then the red blood cells, ultimately leading to the symptoms of malaria.

What Are The Symptoms Of Malaria?

The main symptoms are high fever – caused by the immune system reacting to the protozoa attacking the red blood cells, and anaemia – caused by the resulting destruction of red blood cells. Vomiting and diarrhoea are also common.

The symptoms often show about a week to two weeks after being infected by a mosquito bite, but sometimes it takes much longer, as much as a year or even more.

Malaria is a serious disease that can in some cases be fatal, even when being treated in the West.

How Many People Around The World Get Malaria?

The exact number is not known, but it is thought that hundreds of millions of people a year get malaria in various parts of the world. In 2012 there were thought to be over two hundred million cases of malaria worldwide.

How Many People Die Of Malaria Around The World In A Year?

There are thought to be over half a million deaths a year worldwide due to malaria.

Young children, pregnant women and non-immune travellers from malaria-free areas are particularly vulnerable to the disease when they become infected.

There is some cause for optimism though, as between 2000 and 2015, malaria incidence (the rate of new cases) fell by 37% globally and malaria death rates fell by 60% globally among all age groups.

In the UK, over a thousand people were diagnosed with malaria after returning to the UK from malarial areas of the world in 2012. Two of these people eventually died of malaria.

Where Are People At Risk Of Getting Malaria?

About 3.2 billion people – almost half of the world's population – are at risk of malaria. They live in the tropical regions of the world where the Anopheles mosquito is found. This includes much of Africa (especially the sub-Sahara) and Asia, as well as Central and South America.

Sub-Saharan Africa carries very high share of the global malaria burden. In 2015, the region reported around nine out of ten of malaria cases and malaria deaths worldwide.

In the UK around a thousand travellers a year return with malaria they have contracted in a tropical area.

Why Don't We Vaccinate People Against Malaria?

Vaccination depends on the introduction of a disease-causing pathogen's antigens to the body, so that antibodies against that antigen are produced in the blood – giving the person immunity to that specific disease.

The vaccination process does not work well for malaria for two reasons, firstly the malarial protozoa has many different antigens on its surface which it keeps changing so that the antibodies (which are specific to one antigen) cannot attach to it. Secondly, because the protozoan spends a lot of its lifecycle inside the red blood cells and liver cells it is not exposed to the antibodies, which are found in the blood plasma.

What Efforts Have Been Made To Eliminate Malaria?

Due to the difficulty with vaccination, efforts have concentrated on getting rid of the mosquito vector (using the insecticide DDT) and giving anti-malarial drugs to those already infected.

Neither of these methods has proved effective in the longer term due to both the mosquito and the parasite developed resistance against these chemicals.

What Can Be Done To Prevent People Catching Malaria?

The best way to prevent from catching malaria is to make sure the person is not bitten by the mosquitoes which transmit the disease.

This can be done by covering the body (wearing long-sleeved shirts, collars and trousers, etc) along with the use of insect repellent. Using a mosquito net around the bed (especially one treated with insect repellent) when sleeping, can also help.

However, none of these methods is completely guaranteed to prevent mosquito bites and therefore cannot guarantee prevention from getting malaria. So, it is also advisable to take anti-malarial tablets to prevent getting malaria as well as all the other methods.

How Can Malaria Be Treated?

If malaria is treated quickly and effectively it is possible to make a full recovery.

Which type of medication is used and the treatment will depend on: the type of malaria caught (there are a few different types) and when it was caught, along with how bad the symptoms are.

Section Three: AIDS - A Viral Infectious Disease

What Causes AIDS?

AIDS (Acquired Immune Deficiency Syndrome) is an infectious disease that is caused by the HIV (Human Immunodeficiency Virus) virus. AIDS is not a disease itself, but a syndrome – a collection of different signs and symptoms, including some well-known diseases.

Someone with AIDS has been infected with HIV and as a result has one of a list of diseases such as: tuberculosis, pneumonia or some types of cancer; which are known as the 'AIDS-defining' diseases.

What Are The Symptoms Of AIDS?

AIDS sufferers tend to get AIDS-defining diseases that are not found in normal people because having AIDS causes the person's immune system to be unable to fight infection.

How Does The HIV Virus Cause The Symptoms Of AIDS?

The HIV virus attacks the T cells of the immune system, so the person is vulnerable to getting diseases they would not get otherwise.

What Does HIV+ (HIV Positive) Mean?

HIV+ literally means that the blood test for the antibody in the person's blood has come out positive.

This means that the person has been infected with the HIV virus.

So, HIV+ means the person has HIV.

What Treatments Are Currently Available For AIDS?

Treatments focus on trying to prevent the virus from multiplying and attacking the T cells of the immune system.

Anti-retroviral drugs are used, which inhibit the viral enzyme reverse transcriptase, stopping it from making DNA from RNA.

What Is The Difference Between HIV And AIDS?

HIV is the name of the virus that causes AIDS.

If someone has AIDS they definitely have HIV because this is the virus that causes AIDS, however someone with HIV may well not develop AIDS, largely thanks to medical advances.

For example, in the UK in 2012 only a few hundred people were diagnosed with AIDS; less than half of one percent of the total of people diagnosed as HIV positive.

How Do You Catch HIV?

The HIV virus is found in the body fluids of a person who is infected with HIV; including their semen (if they are male); their vaginal fluids and breast milk (if they are female); and their anal fluids and blood.

So contact between an infected person's body fluids (as listed above) and your own, can result in you getting infected with HIV.

The HIV virus does not survive for long outside the body and cannot be caught from the sweat or urine of an infected person.

What Is The Most Common Way Of Catching HIV?

More than nine out of ten of cases of HIV are thought to have been got through unprotected sexual contact with an infected person, either by anal or vaginal sex.

HIV is also contracted by drug users who share a needle with an infected person or from mother to baby, during pregnancy, birth and breastfeeding.

How Can People Prevent Themselves From Getting HIV?

The surest way for a healthy person to prevent themselves from getting HIV is not to allow contact between their own body fluids and those of an HIV infected person.

This can be done by ensuring a condom is always used when having vaginal or anal sex. It may also be advisable to use a condom during oral sex although there is a much lower risk of transmission of HIV than for vaginal or anal sex.

Needles, syringes or any other injecting equipment should never be shared as this brings blood into contact.

Is There A Cure For HIV?

No, there is no cure.

However, HIV is treatable and can be kept under control so that the immune system stays healthy. This means that people with HIV can often live a healthy, active life – although they may suffer side effects from treatment. The earlier that HIV is diagnosed the more likely treatment is to be effective.

Is It Possible To Have HIV And Not Realise It?

Yes it is possible to have HIV and not realise it, as it may not have any impact at all on your health, for many years, sometimes for ever.

Most people, after being infected with HIV, have a short, flu-like illness a few weeks after infection. After that, these HIV+ people often have no symptoms for several years and appear perfectly healthy.

It is estimated that around a quarter of people who have HIV, don't realise that they have it.

The fact that it cannot often be known by sight if someone has HIV, should of course, be a warning about making sure not to come into sexual or other body fluid contact unknowingly with someone who has HIV (but looks totally healthy) and by doing so put yourself at risk of catching it. It is much better to be safe than sorry.

Section Four: Cholera - A Water-Borne Bacterial Infectious Disease

What Is Cholera?

Cholera is an acute bacterial infection of the digestive system that is carried in contaminated water which contains the bacteria *Vibrio cholerae*.

What Are The Symptoms Of Cholera?

The usual symptoms are severe, watery diarrhoea; along with both feeling and being sick and stomach cramps.
These symptoms can appear very quickly after getting the infection or may more commonly take a few days to develop.

How Serious Is Cholera?

The continual diarrhoea and vomiting can quickly lead to dehydration and cause a rapid and significant drop of blood pressure. This can be fatal, even within a few hours, if left untreated.
Unfortunately, children are particularly vulnerable to this dehydration and tend to suffer badly in cholera epidemics.

Why Does Cholera Mainly Affect People In The Developing World?

This is due to poor sewage and sanitation facilities in the countries of the developing world because cholera infection spreads through drinking water contaminated with the faeces (excrement) of someone who already has the cholera infection, so if sewage and drinking water are not properly separated and sewage not correctly treated before it is discharged, as in most of the developing world; there is nothing to prevent the disease from occurring.

The countries where there is the greatest risk of getting cholera are: south and south-east Asia, parts of the Middle East and sub-Saharan Africa. Within these countries those areas with poor sanitation are most at risk. Refugee camps are also at risk for the same reason of poor sanitation and difficulty in obtaining clean drinking water.

It should not be forgotten that terrible, devastating cholera epidemics occurred historically in London and other cities in Europe when there was poor sanitation, such as during the 18th and early 19th century.

Cholera is preventable everywhere, provided that there is clean drinking water and proper sanitation.

How Is Cholera Treated?

As the main symptoms of diarrhoea and vomiting can cause severe dehydration; which itself can lead to shock due to low blood pressure; so the most common treatment is oral re-hydration therapy (ORT), which helps to replace lost fluid and salts. About eighty percent of those given ORT recover successfully from cholera.
However, severely cases may need intravenous fluids and antibiotics to reduce diarrhea and kill the infecting cholera bacteria.

How Many People Get Cholera Worldwide Per Year?

There are thought to be around four million cases of cholera per year worldwide

How Many People Die Of Cholera Worldwide Per Year?

There is estimated to be around one hundred thousand deaths per year worldwide from cholera.

Is It Possible To Vaccinate Against Cholera?

Yes it is possible.

The cholera vaccine is easily and painlessly taken as a drink, taken a few times over the course of a month or so. It is about 80% effective, but its effect reduces over a period of months and a booster is usually needed.

Section Five: Tuberculosis (TB), a Droplet-Infection Bacterial Disease

What Is The Historical Background To TB?

Tuberculosis has been known to mankind since ancient times. Earlier this disease has been called by numerous names including consumption (because of severe weight loss) and the white plague (because of the pale skin colouring seen among sufferers).

The organism that causes tuberculosis - Mycobacterium tuberculosis is known to have existed 20,000 years ago. It has been found in relics from ancient Egypt, India, and China.

In the nineteenth century, tuberculosis was known as "the captain of all men of death". Sadly, to some extent, there is still some truth in that, even in our times.

What Is The Current Worldwide Situation With TB?

Even today after the development of advanced screening, diagnostic and treatment methods for the disease, a third of the world's population has been exposed and is infected with the organism that causes TB. It is thought that number is as high as nine out of ten people in the developing world.

Since the appearance of HIV infection in the 1980s there has been a huge resurgence of TB with more than eight million new cases each year worldwide and more than two million deaths as a result.

It's estimated around one-third of the world's population is infected with latent TB (meaning that they have the bacteria in their bodies but have no symptoms of the infection at all). Of these, up to 10% will become active (meaning that they will get the symptoms of TB) at some point.

What Causes TB?

Infection by two types of bacteria can cause TB: Mycobacterium tuberculosis or Mycobacterium bovis; the latter is found in cows.

In most healthy people the immune system (the body's natural defence against infection and illness) kills the bacteria and you have no symptoms.

Sometimes the immune system cannot kill the bacteria, but manages to prevent it spreading in the body. This means you will not have any symptoms, but the bacteria will remain in your body. This is known as latent TB.

If the immune system fails to kill or contain the infection, it can spread within the lungs or other parts of the body and symptoms will develop within a few weeks or months. This is known as active TB.

Latent TB could develop into an active TB infection at a later date, particularly if your immune system becomes weakened.

How Do You Catch TB?

Tuberculosis (TB) is a bacterial infection spread through inhaling tiny droplets from the coughs or sneezes of an infected person. So, TB is spread from person to person through the air. When people with lung TB cough, sneeze or spit, they propel the TB germs into the air.

TB that affects the lungs is the most contagious type, but it usually only spreads after prolonged exposure to someone with the illness. For example, it often spreads within a family who live in the same house.

How Common Are Antibiotic Resistant Strains Of TB?

Globally in 2014, it was estimated that around half a million people developed multi-drug-resistant TB (MDR-TB).

What Are The Symptoms of TB?

As TB usually takes the form of an active lung infection its typical symptoms include:

- a lasting cough, bringing up phlegm, some of which may be bloody
- a high body temperature
- tiredness and fatigue
- loss of appetite and body weight
- new swellings that haven't gone away after a few weeks

How Many People Catch and Die from TB, Worldwide?

Tuberculosis (TB) is one of the diseases that kill the most people worldwide.

In 2014, it is estimated that almost ten million people fell ill with TB, of these an estimated 1 million were children who became ill with TB

In 2014, about one and a half million people died from it; and 140,000 children died of TB.

In Which Parts of the World Is TB Most Common?

More than nine out of ten of TB deaths occur in low- and middle-income countries such as the Third World.

In 2014, about eight out of ten of reported TB cases occurred in just over twenty countries. The six countries having the largest number of TB cases were India, Pakistan, Indonesia, Nigeria, The People's Republic of China and South Africa.

Is TB Still Found In The UK?

There are cases of TB still in the UK. In 2013 around 8,000 cases of TB were reported in the UK. Of these, the majority of sufferers were people who were born outside the UK

Before antibiotics and vaccination were introduced around the time of the Second World War, TB was a common health problem in the UK. Nowadays, the condition is much less common, but in the last 20 years TB cases have gradually increased, particularly due to immigration of people are originally from places where TB is more common.

Is It Possible To Vaccinate Against TB?

There is a vaccine against TB, called the BCG, which provides effective protection against TB in around four out of five people who are given it.

Currently, BCG vaccinations are only recommended for groups of people who are at a higher risk of developing TB.

This includes children living in areas with high rates of TB, or those who have close family members from countries with high TB rates, and people under the age of 16 who are going to live and work with local people in an area with high rates of TB for more than three months.

It's also recommended that some people, such as healthcare workers, are vaccinated because of the increased risk of contracting TB while working.

Chapter Three: Exam Questions and Answers

Question 1
Name the pathogens that cause the following diseases: cholera and tuberculosis. (2 marks)

Exam Tip: this is very straightforward, just a matter of remembering which microbe causes the disease.

The Answer: both cholera and tuberculosis are caused by bacteria; cholera is caused by Vibrio cholerae (1) and tuberculosis is caused by Mycobacterium tuberculosis or Mycobacterium bovis (2).

Question 2
How are AIDS and malaria transmitted and how can this be prevented? (2 Marks)

Exam Tip: remember to mention transmission and prevention for both AIDS and malaria.

The Answer: AIDS is transmitted by contact with the body fluids of an infected person and this can be prevented by using condoms and not sharing needles (1). Malaria is transmitted by mosquito bites and this can be prevented by killing mosquitoes with insecticide and using mosquito nets when in bed (2).

www.ingramcontent.com/pod-product-compliance
Lightning Source LLC
Chambersburg PA
CBHW070428190526
45169CB00003B/1465